The Quotable
FARM WIFE

Norvia Behling

Voyageur Press

First published in 2007 by Voyageur Press, an imprint of MBI Publishing Company, Galtier Plaza, Suite 200, 380 Jackson Street, St. Paul, MN 55101 USA

Copyright © 2007 by Norvia Behling

All rights reserved. With the exception of quoting brief passages for the purposes of review, no part of this publication may be reproduced without prior written permission from the Publisher.

The information in this book is true and complete to the best of our knowledge. All recommendations are made without any guarantee on the part of the author or Publisher, who also disclaim any liability incurred in connection with the use of this data or specific details.

We recognize, further, that some words, model names, and designations mentioned herein are the property of the trademark holder. We use them for identification purposes only. This is not an official publication.

Voyageur Press titles are also available at discounts in bulk quantity for industrial or sales-promotional use. For details write to Special Sales Manager at MBI Publishing Company, Galtier Plaza, Suite 200, 380 Jackson Street, St. Paul, MN 55101 USA.

To find out more about our books, join us online at www.VoyageurPress.com.

All photography by Norvia Behling except on pages 1, 2, 5, 7, 11, 12, 16, 18, 21, 24, 28, 32, 35, 36, 39, 44, 46, 49, 50, 52, 55, 61, 62, 64, 68, 71, 74, 77, 78, 80, 87, 88, 92, and the back cover by Daniel Johnson and pages 30, 40, 42, 56, 67, 72, 82, and 94 by Paulette Johnson.

Editor: Amy Glaser
Designer: Sara Holle

Printed in China

Library of Congress Cataloging-in-Publication Data

The quotable farm wife / [compiled] by Norvia Behling.
  p. cm.
"Editor: Amy Glaser."
 ISBN-13: 978-0-7603-2910-8 (hardbound w/ jacket)
 ISBN-10: 0-7603-2910-9 (hardbound w/ jacket)
1. Women in agriculture—Quotations, maxims, etc.
2. Farmers' spouses—Quotations, maxims, etc.
3. Farm life—Quotations, maxims, etc. I. Behling, Norvia. II. Glaser, Amy.
PN6084.W6Q675 2007
630.82—dc22
                                2007002671

Farmers are the *salt* of the *earth*.

—⋅—

Anonymous

A *day away* from some *people* is

like a *month* in the *country*.

—·⁂·—

Howard Dietz

To tell a *woman* everything she may *not do*

is to tell her *what* she *can do*.

※

Spanish Proverb

Summer has *set* in with its usual *severity*.

Samuel Taylor Coleridge

Those who *never* sink into this *peace* of *nature* lose a *tremendous* well of *strength*, for there is something *healing* and *life-giving* in the mere *atmosphere* surrounding a *country* house.

—⁂—

Eleanor Roosevelt

Never send a *boy* to do a

man's job—send a *woman*.

⁃

Anonymous

Take *rest*; a field that has *rested* gives a *bountiful* crop.

—⊢—

Ovid

There can be no *progress* of *people*

who have no *faith* in tomorrow.

⁃┼⁃

John F. Kennedy

He who marries a *wife* reared on the *land*

marries *strength* and *purity* and *compassion*.

—I—

Henry Ward Beecher

Another *difference* between the *rich* and the *poor* is that the *rich* are tired in the morning and the *poor* are tired in the *evening*.

Anonymous

Carpe diem—*seize* the day.

—✠—

Horace

Even after a *bad* harvest there must be *sowing*.

—✠—

Lucius Annaeus Seneca

When you have a plot of *land*, the best thing to *cultivate* is your *neighbors*.

Anonymous

There are only *two* ways to be contented:

one is *liking* what you *do* and the

other is *doing* what you *like*.

—✦—

Anonymous

The best *preparation* for good work

tomorrow is to do good work *today*.

—☙—

Elbert Hubbard

There is *no* substitute for *hard* work.

—⁂—

Thomas A. Edison

Winter is not a *season*,

it's an *occupation*.

— Sinclair Lewis

Nothing is got without *pain* but dirt and long nails.

—⊹—

English Proverb

The *people* who get on in this world are the people who get up and *look* for the circumstances they *want*, and if they *can't* find them, *make* them.

—◆—

George Bernard Shaw

Reduce the *complexity* of life by eliminating

the needless *wants* of life, and the

labors of life *reduce* themselves.

—I—

Edwin Way Teale

You've got to *stop* and *smell* the roses.

—⊹—

Anonymous

We shape our *dwellings*, and afterwards

our *dwellings* shape *us*.

—I—

Winston Churchill

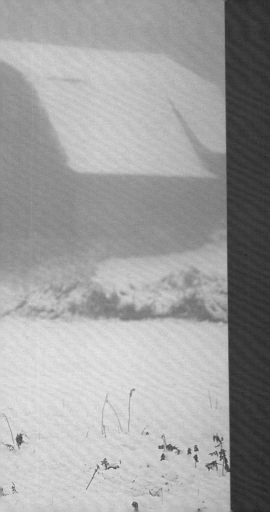

Woman is a being—

therefore *let her* be.

Anonymous

The *secret* of success is this:

there is *no* secret of success.

⁃✦⁃

Elbert Hubbard

We look *too* much to museums.

The *sun* coming up in the morning is *enough*.

—⁂—

Romare Bearden

Things turn out *best* for people who *make*

the best of the *way* things turn *out*.

—◆—

John Wooden

Anybody can be *good* in the *country*.

—✢—

Oscar Wilde

Strengthen me by *sympathizing* with

my *strength, not* my weakness.

—⧓—

Amos Bronson Alcott

Wildness is a bench mark, a *touchstone*.

In the wilderness we can *see* where we have come

from, *where* we are going, how *far* we've gone.

In *wilderness* is the only unsullied earth sample

of the forces generally at *work* in the universe.

Kenneth Bower

Not *knowing* when the dawn

will come, I open every door.

⊷I⊶

Emily Dickinson

Water, thou hast no taste, no color, no odor; canst not be *defined*, art relished while ever *mysterious*. Not necessary to life, but rather life *itself*, thou fillest us with a gratification that exceeds the *delight* of the senses.

—✛—

Antoine de Saint-Exupery

Take a *lesson* from spring: say it with *flowers.*

—⊹—

Anonymous

We won't have a *society* if we *destroy* the *environment*.

⁂

Margaret Mead

Nothing can come of *nothing*.

—⊹—

William Shakespeare

A woman uses her *intelligence* to find reasons to support her *intuition.*

— I —

Gilbert K. Chesterton

It is thus with *farming*; If you do one thing *late*, you will be late in *all* your work.

—⁂—

Cato the Elder

It makes all the *difference*

whether the shepherd *loves*

the *fleece* or the *flock*.

—✥—

Anonymous

Women are *not* the weak, frail little flowers that they are advertised. There has *never* been anything invented yet, including *war*, that a man *would* enter into, that a woman *wouldn't*, too.

Will Rogers

The *woman* who knows *all* the answers never gets *asked*.

—⁃—

Anonymous

We do not remember *days*, we remember *moments*.

―✢―

Cesare Pavese

We *never* reflect how *pleasant* it is to ask for *nothing*.

⋅┼⋅

Lucius Annaeus Seneca

A woman is like a *teabag*; you never know how *strong* she is until she gets in *hot water*.

—I—

Eleanor Roosevelt

The *wife* of Atlas probably *supported*

him while he supported the *world*.

※

Anonymous

Our *life* is what our *thoughts* make it.

—⋈—

Marcus Aurelius

The *city* has a *face*, the *country* a *soul*.

—✦—

Jacques de Lacretelle

About the Photographers
Norvia Behling's photographs have appeared thousands of times in books, magazines, and calendars. She has been capturing the unique qualities of her subjects for 25 years. She divides her time between her farm in Wisconsin in the summers and her home in Florida in the winters. Norvia is married and has one daughter.

Daniel Johnson specializes in equine photography, but he also enjoys photographing many other subjects, such as dogs, farm animals, gardens, and rural life. His images are found in magazines, books, greeting cards, and calendars nationwide. Dan also manages the family owned horse farm and oversees the breeding, training, and showing of their horses. Two of his books include *The World's Greatest Horse Poster Book* and *How To Raise Horses*.

Paulette Johnson lives with her husband and their five children on a farm in northern Wisconsin where they raise Welsh ponies. Paulette works closely with her sister, Norvia Behling, in their photography pursuits. Paulette enjoys photographing farm life and horses.